THE SCIENCE OF ANIMAL MOVEMENT

How Mammals Run

BY EMMA HUDDLESTON

CONTENT CONSULTANT
DAVID HU, PHD
PROFESSOR
MECHANICAL ENGINEERING
GEORGIA TECH

Kids Core
An Imprint of Abdo Publishing
abdobooks.com

abdobooks.com

Published by Abdo Publishing, a division of ABDO, PO Box 398166, Minneapolis, Minnesota 55439. Copyright © 2021 by Abdo Consulting Group, Inc. International copyrights reserved in all countries. No part of this book may be reproduced in any form without written permission from the publisher. Kids Core™ is a trademark and logo of Abdo Publishing.

Printed in the United States of America, North Mankato, Minnesota
042020
092020

THIS BOOK CONTAINS RECYCLED MATERIALS

Cover Photo: Stu Ports/iStockphoto
Interior Photos: iStockphoto, 4–5; Frank Hildebrand/iStockphoto, 6; Shutterstock Images, 9, 12–13, 14, 18–19, 25 (bottom), 28, 29; Michelle Lalancette/Shutterstock Images, 10; Rostislav Stach/Shutterstock Images, 16; Dennis W. Donohue/Shutterstock Images, 21; Paula French/Shutterstock Images, 22; TW Photos/iStockphoto, 23; Irina Maksimova/Shutterstock Images, 25 (top)

Editor: Marie Pearson
Series Designer: Ryan Gale

Library of Congress Control Number: 2019954243

Publisher's Cataloging-in-Publication Data

Names: Huddleston, Emma, author.
Title: How mammals run / by Emma Huddleston
Description: Minneapolis, Minnesota : Abdo Publishing, 2021 | Series: The science of animal movement | Includes online resources and index.
Identifiers: ISBN 9781532192968 (lib. bdg.) | ISBN 9781644944356 (pbk.) | ISBN 9781098210861 (ebook)
Subjects: LCSH: Children's questions and answers--Juvenile literature. | Mammals--Behavior--Juvenile literature. | Science--Examinations, questions, etc--Juvenile literature. | Habits and behavior--Juvenile literature.
Classification: DDC 500--dc23

CONTENTS

CHAPTER 1
A Cheetah Running 4

CHAPTER 2
Energy 12

CHAPTER 3
Cycle of Steps 18

Movement Diagram 28
Glossary 30
Online Resources 31
Learn More 31
Index 32
About the Author 32

Cheetahs need to be fast because their prey is fast.

CHAPTER 1

A Cheetah Running

A flash of tan and black spots crosses the open field in Kenya. A cheetah chases its prey. Its front legs reach forward. Its strong back legs push off the ground. A cheetah's legs make its body spring forward.

Pronghorns reach speeds up to 55 miles per hour (89 km/h).

Cheetahs are the fastest running mammal. They can run in bursts at more than 70 miles per hour (110 km/h). But their speed usually lasts less than 60 seconds. In North America, pronghorn antelopes are not as fast as cheetahs. But they can run long distances faster than any other mammal.

Moving Their Muscles

Mammals live in every habitat. They walk, climb, swim, and crawl. Land mammals can weigh less than 1 pound (0.5 kg) or more than 7,000 pounds (3,200 kg). Some mammals have two legs. Others have four. Mammals with four legs are usually faster runners.

Mammals use muscle movement to get around. They use energy to tighten their muscles. Once tightened, the muscles pull or push on the bones they are connected to.

Stride and Speed

A mammal's speed is determined by its stride length and stride rate. Stride length is the distance of one step. Ungulates have long stride lengths. These are mammals that have hooves, such as deer and horses. Many have long legs that are great for running.

No Collarbone

Mammals that have adapted to run don't have a collarbone. This helps them move faster. Their shoulders can easily reach forward. Without a collarbone, their shoulders add to the stride length.

Long legs help red deer run fast.

Wolves can move their legs quickly, which helps them run fast.

Stride rate is about how quickly a mammal takes steps. Carnivores such as wolves tend to have shorter and stockier legs. They use their legs to catch prey as well as run. Shorter legs have a short stride length, but they can take steps more quickly. The science behind a mammal's muscles and steps explains how it runs.

Explore Online

Visit the website below. Did you learn any new information about mammals that wasn't in Chapter One?

Mammal Pictures & Facts

abdocorelibrary.com/how-mammals-run

Giraffes run slower than smaller animals such as gazelles.

CHAPTER 2

Energy

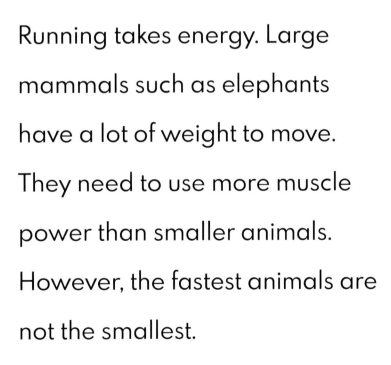

Running takes energy. Large mammals such as elephants have a lot of weight to move. They need to use more muscle power than smaller animals. However, the fastest animals are not the smallest.

Wildebeests are among the fastest runners. They can reach 50 miles per hour (80 km/h).

Small animals have less weight, but they also have smaller muscles. They have limited muscle power. The fastest mammals are midsized. Cheetahs and jaguars are midsized. They have enough muscle to balance out their weight.

The faster an animal runs, the more energy it uses. Mammals must overcome inertia to run. Inertia is the idea that an object won't start or stop moving unless an outside force makes it do so. For example, a rock will not move unless someone or something uses energy to move it. To start moving, animals must use energy from their muscles to power their movement. This energy mainly comes from rear legs when they push off the ground.

Speedy Rabbits

Rabbits have powerful back legs. They push off the ground to hop or run. Jackrabbits are the fastest rabbits. They can move at speeds up to 45 miles per hour (70 km/h).

Predators use fast-twitch muscles to chase down prey.

Different Muscles

Mammals have different types of muscles for different kinds of running. Running fast over short distances requires fast-twitch muscles.

These muscles move quickly. They allow an animal to take many steps rapidly.

Slow-twitch muscles are helpful for running longer distances at slower speeds. These muscles don't move as quickly. Since they help animals run at slower speeds, they tend to use up energy slower.

Further Evidence

Look at the website below. Does it give any new evidence to support Chapter Two?

Motion

abdocorelibrary.com/how-mammals-run

Gazelles need speed to escape predators.

CHAPTER **3**

Cycle of Steps

When mammals run, they take step after step. There are two parts to the cycle of a step. The first half of the cycle starts with footfall. Footfall is when a foot touches down onto the ground. The first half ends when that foot lifts off the ground.

During this part of the cycle, the foot and leg don't move much. They are pushing against the ground. This moves the body forward. When the foot first hits the ground, the body is behind the foot and leg. Then it passes over the foot and leg. By the time the foot lifts off, the body is in front of the foot and leg.

The second half of the cycle begins with liftoff. It ends with another footfall. It happens right after the first half of the cycle and leads to the start of a new step cycle. During this half of the cycle, the body is mostly still. Muscles move the foot and leg forward.

An animal's foot stays planted on the ground as its leg pushes the body forward.

While galloping, all four legs gather underneath the animal.

Galloping

Galloping is the fastest way mammals with four legs run. Each leg moves in a pattern of steps. Four steps equal one stride. Most four-legged

When animals gallop, all four feet hit the ground at a different time.

mammals have a transverse gallop. In the transverse gallop, there is one suspension. A suspension is time spent with no feet on the ground. This happens when the feet are all gathered under the body.

Certain ungulates, including horses, gallop by mainly moving their legs. They have rigid spines. Their legs create nearly all of their stride length.

Some mammals also have a rotary gallop. This type of gallop has two suspension parts. The first is similar to the regular gallop. It happens when all feet are gathered under the body. The second is when the front feet reach out in front and the rear feet extend behind the body. The rotary gallop is especially common for carnivores, including wolves and cheetahs. Some ungulates, including gazelles, also use it.

Spine Made for Speed

Greyhounds are dogs that are built for a rotary gallop. At top speed, their feet touch the ground only 25 percent of the time. A greyhound's deep chest holds its large heart. This heart pumps blood around its body, sending oxygen to its muscles.

Transverse and Rotary Galloping

Transverse Gallop

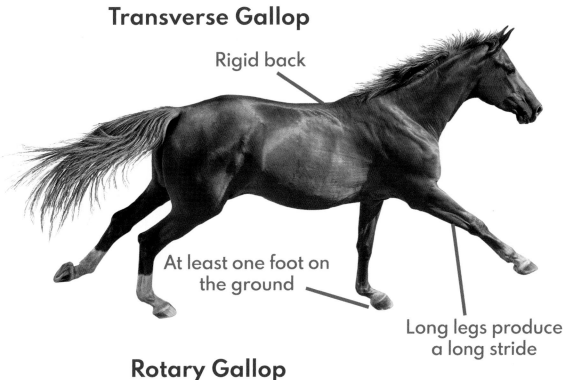

- Rigid back
- At least one foot on the ground
- Long legs produce a long stride

Rotary Gallop

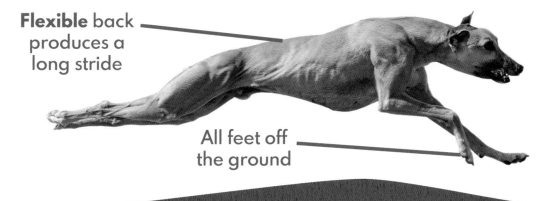

- **Flexible** back produces a long stride
- All feet off the ground

This image compares a horse using the transverse gallop to a dog using the rotary gallop. When the legs are fully extended in the transverse gallop, at least one foot is still on the ground. But in the rotary gallop, all feet are off the ground.

Mammals that use a rotary gallop have flexible spines. They push off the ground with their hind feet. Then they stretch their spines to reach out far. This increases their stride length. But this type of running takes lots of energy. For that reason, carnivores often run fast in bursts. Other mammals can run faster for longer periods of time.

No matter how a mammal runs, it uses muscle power. Different types of muscles help mammals move their legs. Different sizes and shapes of legs help mammals take step after step.

Primary Source

Alan Wilson studies animal movement. He explains how prey can escape hunters that are faster, such as cheetahs:

> The [best move] of the prey is to run relatively slowly and turn very sharply at the last moment. . . .
>
> [Prey] decides when to turn . . . so it's always one stride ahead of the predator.

Source: Helen Briggs. "How to Escape from a Lion or Cheetah—the Science." *BBC News*, 24 Jan. 2018, bbc.com. Accessed 18 Dec. 2019.

Comparing Texts

Does this quote support the information in this chapter? Or does it give a different perspective? Explain how in two to three sentences.

Movement Diagram

Transverse Gallop

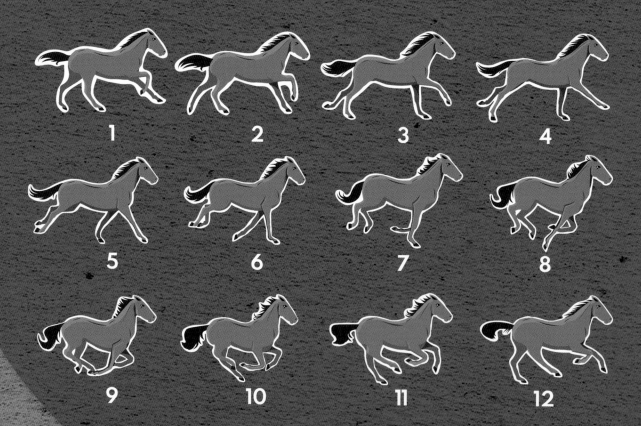

Rotary Gallop

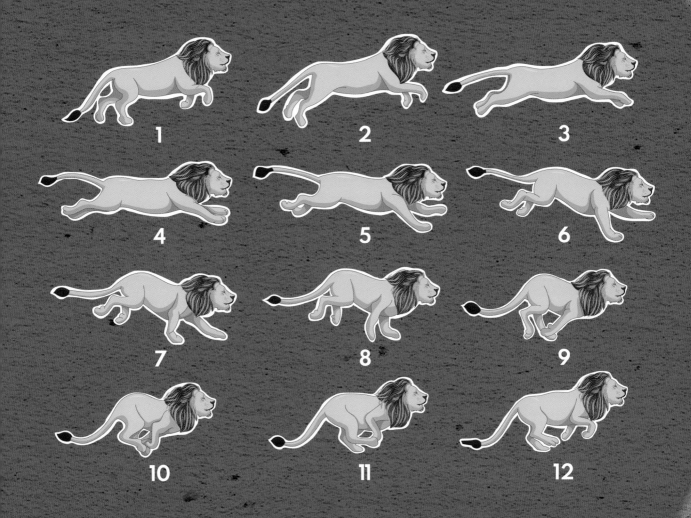

Glossary

carnivores
animals that eat meat as the main part of their diet

flexible
being able to bend, stretch, and move in multiple directions

muscle
a part of the body that can tighten in order to make the body move

prey
an animal that is hunted by another

rigid
stiff, hard, or unable to bend much

ungulates
mammals that eat plants and have hooves

Online Resources

To learn more about how mammals run, visit our free resource websites below.

Visit **abdocorelibrary.com** or scan this QR code for free Common Core resources for teachers and students, including vetted activities, multimedia, and booklinks, for deeper subject comprehension.

Visit **abdobooklinks.com** or scan this QR code for free additional online weblinks for further learning. These links are routinely monitored and updated to provide the most current information available.

Learn More

DePrisco, Dorothea. *Animals on the Move.* Liberty Street, 2017.

Murray, Julie. *Mammals.* Abdo Publishing, 2019.

Watson, Galadriel. *Running Wild.* Annick Press, 2020.

Index

carnivores, 11, 24, 26
cheetahs, 5–7, 14, 24, 27

energy, 7, 13, 15, 17, 26

greyhounds, 24

horses, 8, 23, 25

inertia, 15

Kenya, 5

muscles, 7, 11, 13–17, 20, 24, 26

pronghorn antelopes, 7

rabbits, 15
rotary gallop, 24–26

step cycle, 19–20
stride length, 8, 11, 23, 26
stride rate, 8, 11

transverse gallop, 23, 25

ungulates, 8, 23–24

Wilson, Alan, 27
wolves, 11, 24

About the Author

Emma Huddleston lives in the Twin Cities with her husband. She enjoys reading, writing, and swing dancing. She thinks the science of animal movement is fascinating!